M336
Mathematics and Computing: a third-level course

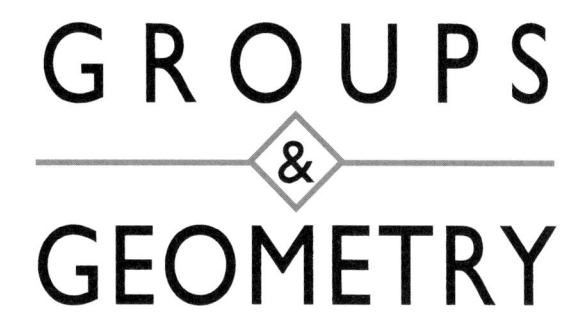

GROUPS
& GEOMETRY

UNIT GR5
SYLOW'S THEOREMS

Prepared for the course team by
Bob Coates & Bob Margolis

The Open University

This text forms part of an Open University third-level course.
The main printed materials for this course are as follows.

Block 1
Unit IB1 Tilings
Unit IB2 Groups: properties and examples
Unit IB3 Frieze patterns
Unit IB4 Groups: axioms and their consequences

Block 2
Unit GR1 Properties of the integers
Unit GR2 Abelian and cyclic groups
Unit GE1 Counting with groups
Unit GE2 Periodic and transitive tilings

Block 3
Unit GR3 Decomposition of Abelian groups
Unit GR4 Finite groups 1
Unit GE3 Two-dimensional lattices
Unit GE4 Wallpaper patterns

Block 4
Unit GR5 Sylow's theorems
Unit GR6 Finite groups 2
Unit GE5 Groups and solids in three dimensions
Unit GE6 Three-dimensional lattices and polyhedra

The course was produced by the following team:

Andrew Adamyk (BBC Producer)
David Asche (Author, Software and Video)
Jenny Chalmers (Publishing Editor)
Bob Coates (Author)
Sarah Crompton (Graphic Designer)
David Crowe (Author and Video)
Margaret Crowe (Course Manager)
Alison George (Graphic Artist)
Derek Goldrei (Groups Exercises and Assessment)
Fred Holroyd (Chair, Author, Video and Academic Editor)
Jack Koumi (BBC Producer)
Tim Lister (Geometry Exercises and Assessment)
Roger Lowry (Publishing Editor)
Bob Margolis (Author)
Roy Nelson (Author and Video)
Joe Rooney (Author and Video)
Peter Strain-Clark (Author and Video)
Pip Surgey (BBC Producer)

With valuable assistance from:

Maths Faculty Course Materials Production Unit
Christine Bestavachvili (Video Presenter)
Ian Brodie (Reader)
Andrew Brown (Reader)
Judith Daniels (Video Presenter)
Kathleen Gilmartin (Video Presenter)
Liz Scott (Reader)
Heidi Wilson (Reader)
Robin Wilson (Reader)

The external assessor was:

Norman Biggs (Professor of Mathematics, LSE)

The Open University, Walton Hall, Milton Keynes, MK7 6AA.

First published 1994. Reprinted 1997, 2002, 2003, 2007.

Copyright © 1994 The Open University

All rights reserved. No part of this publication may be reproduced, stored in a retrieval system or transmitted in any form or by any means, without written permission from the publisher or a licence from the Copyright Licensing Agency Limited. Details of such licences (for reprographic reproduction) may be obtained from the Copyright Licensing Agency Ltd of 90 Tottenham Court Road, London, W1P 9HE.

Edited, designed and typeset by the Open University using the Open University TEX System.

Printed in Malta by Gutenberg Press Limited.

ISBN 0 7492 2167 4

This text forms part of an Open University Third Level Course. If you would like a copy of *Studying with the Open University*, please write to the Central Enquiry Service, PO Box 200, The Open University, Walton Hall, Milton Keynes, MK7 6YZ. If you have not already enrolled on the Course and would like to buy this or other Open University material, please write to Open University Educational Enterprises Ltd, 12 Cofferidge Close, Stony Stratford, Milton Keynes, MK11 1BY, United Kingdom.

1.4

CONTENTS

Study guide	4
Introduction	5
1 Sylow p-subgroups	6
2 Sylow's first and second theorems	13
3 Sylow's third theorem (audio-tape section)	19
4 Applications of the Sylow theorems	25
5 Subgroups of prime power order	29
Solutions to the exercises	33
Objectives	44
Index	44

STUDY GUIDE

The first three sections of this unit are of roughly equal length, in terms of study time. Section 4, consisting mainly of exercises, will probably take you considerably longer. The last section, in compensation, is rather shorter.

The theorems proved in this unit are powerful, general ones. It follows that some of the work, that in Sections 2 and 5 particularly, is fairly theoretical. The exercises in these sections are mainly restricted to providing some of the steps in various proofs. The proofs themselves are included because of their importance in the development of the theory of finite groups. For this course, it is the application of the theorems to the determination of the structure of particular groups that is important; so the other sections apply the theorems to particular groups, and the exercises in those sections are often less abstract.

There is an audio programme associated with Section 3.

INTRODUCTION

In this unit we continue our investigation of finite groups.

We know, from Section 3 of *Unit GR4*, that a finite group may fail to have a subgroup corresponding to some divisors of its order. For example, the alternating group A_4 of order 12 has no subgroup of order 6. However, we shall show that there is a subgroup corresponding to every *prime power* divisor of the order of the group. This theorem will be the main result of this unit. Thus, for example, this theorem will ensure that a group of order

$$24 = 2^3 \times 3$$

has subgroups of orders 2, 2^2, 2^3 and 3.

We have already shown, in *Unit GR4*, that this is true for p-groups.

Any subgroup whose order is the highest power of a particular prime dividing the order of the group has additional properties that can be of assistance in classifying the groups of a given order. For example, the theorems that we shall prove will enable us to say that any group of order $91 = 7 \times 13$ must have a normal subgroup of order 13. This, in turn, will enable us to show that any group of order 91 is cyclic.

For a group of order 24, 2^3 is the highest power of 2 dividing the order.

Broadly speaking, the results that we shall prove guarantee the existence of subgroups corresponding to prime power divisors of the order of a group and provide restrictions on the number of such subgroups a group can possess.

The classic proofs of these results, which are due to the Norwegian mathematician L. Sylow, start by considering the special cases for Abelian groups. However, the tools provided by group actions give more concise proofs, and these are the proofs that we shall give.

The name Sylow is usually pronounced as 'Siloff' or 'Silow', although we are informed that the correct pronunciation is 'Sooloff'.

In Section 1, we introduce the first of the new group actions and investigate its application to some particular examples. These examples will point the way to the general results that are needed. The use of group actions requires us to know the number of elements in the set being acted on. In the actions that we use, this involves knowledge of certain binomial coefficients.

Section 2 generalizes the special cases of Section 1 to provide a proof of the first two of Sylow's theorems.

Section 3 completes the proofs of Sylow's results with a proof of his third theorem.

In Section 4, we apply the results obtained to groups of various orders.

Finally, Section 5 shows that some of the results generalize to any prime power divisor of the order of a group.

1 SYLOW p-SUBGROUPS

The main aim of this unit is to establish the Sylow theorems, the first of which is as follows.

Theorem 1.1 Sylow's first theorem

Let G be a finite group of order n and let p be a prime dividing n, where p^α is the highest power of p dividing n.
Then G has a subgroup of order p^α.

Such a subgroup is called a **Sylow p-subgroup** of G.

The divisor p^α of n which is the largest power of p that divides n is referred to as the **maximal prime power divisor** corresponding to p.

Thus, for example, the theorem tells us that the permutation group S_4, which has order $4! = 24$, has subgroups of orders $2^3 = 8$ and 3.

From *Unit GR4*, we know that a p-group has subgroups corresponding to every divisor of its order. Taken together with Sylow's First Theorem, this means that a finite group has a subgroup corresponding to *every* prime power divisor of its order. As this is sometimes useful, we state it as a corollary.

Theorem 5.1 of Unit GR4.

Corollary 1.1

Let G be a finite group of order n and let p be a prime dividing n.
If p^β divides n then G has a subgroup of order p^β.

Hence, the permutation group S_4 also has subgroups of orders 2 and $2^2 = 4$ as well as the ones mentioned above.

In Example 1.1 we illustrate the structure of the proof of Sylow's First Theorem by considering a specific example. In the example, we shall introduce the group action that we shall later use to prove the theorem. The group action we want is a generalization of the action of a group on itself defined by left multiplication.

The proof itself is postponed until Section 2.

Example 1.1

Let G be any group of order $20 = 2^2 \times 5$. The order of G is divisible by the primes 2 and 5. The maximal prime power divisors of $|G|$ corresponding to 2 and 5 are $2^2 = 4$ and $5^1 = 5$, respectively.

We shall show that G must have a subgroup of order 4.

We define a group action as follows. The group acting is G itself. We now define the set X on which G acts to be the collection of *all* 4-element subsets of G.

Next, we define the action of the group G on the set X. It is defined by left multiplication:
$$g \wedge A = gA = \{ga : a \in A\}$$
for every element A of X.

In order to verify that this really is a group action, we must show that, for all $g, h \in G$ and $A \in X$:

(a) $g \wedge A$ is an element of X, i.e. it is a 4-element subset of G;
(b) $e \wedge A = A$;
(c) $(gh) \wedge A = g \wedge (h \wedge A)$.

For most actions, the fact that $g \wedge A$ is an element of the set X is sufficiently obvious that it does not warrant a proof. For our present action we do not feel justified in taking this for granted.

So that we do not break the flow of the example, we shall not do this here. We shall ask you do the verification for a general group in Exercise 1.2.

As for any group action, we can immediately make two statements. One comes from the partition equation, i.e. from the fact that the orbits partition X; the other comes from the Orbit–stabilizer Theorem.

Suppose that
$$\mathrm{Orb}(A_1), \mathrm{Orb}(A_2), \ldots, \mathrm{Orb}(A_s)$$
are the distinct orbits for this group action. Then, by the partition equation,
$$|X| = |\mathrm{Orb}(A_1)| + |\mathrm{Orb}(A_2)| + \cdots + |\mathrm{Orb}(A_s)|$$
and, by the Orbit–stabilizer Theorem,
$$|G| = |\mathrm{Orb}(A_i)| \times |\mathrm{Stab}(A_i)|, \quad i = 1, \ldots, s.$$

Since we are searching for a subgroup of order 4, we have two possible lines of attack. We could show that one of the 4-element subsets that are the elements of X must be a subgroup of G. Alternatively, we could show that one of the stabilizers, which we *know* are subgroups of G, must have 4 elements. Both approaches are possible; we have chosen the second.

To make use of the equations above, we need to know $|X|$, the size of X. Since G has 20 elements, the number of 4-element subsets of G is the number of ways of choosing 4 elements from 20, that is
$$\begin{aligned} {}^{20}C_4 &= \frac{20 \times 19 \times 18 \times 17}{4 \times 3 \times 2 \times 1} \\ &= \frac{5 \times 19 \times 9 \times 17}{1 \times 3 \times 1 \times 1} \\ &= 5 \times 19 \times 3 \times 17 \\ &= 4845. \end{aligned}$$

The *binomial coefficient* ${}^n C_r$ can be calculated from
$$\begin{aligned} {}^n C_r &= \frac{n!}{r!(n-r)!} \\ &= \frac{n \times (n-1) \times \cdots \times (n-r+1)}{r!} \end{aligned}$$

Hence, $|X| = 4845$.

Thus, the partition equation is
$$4845 = |\mathrm{Orb}(A_1)| + |\mathrm{Orb}(A_2)| + \cdots + |\mathrm{Orb}(A_s)|$$
and, from the Orbit–stabilizer Theorem, we have
$$20 = |\mathrm{Orb}(A_i)| \times |\mathrm{Stab}(A_i)|, \quad i = 1, \ldots, s.$$

We shall use these equations to deduce that there is a stabilizer whose order is 4. For such a stabilizer, by the Orbit–stabilizer Theorem, the size of the corresponding orbit is 5, and hence is *not divisible by* 2. This observation prompts us to look for an orbit whose size is not divisible by 2.

Now, 4845 is not divisible by 2, and the partition equation is
$$4845 = |\mathrm{Orb}(A_1)| + |\mathrm{Orb}(A_2)| + \cdots + |\mathrm{Orb}(A_s)|.$$

If all the terms in the sum on the right-hand side were divisible by 2, then so would the left-hand side be. But the left-hand side is *not* divisible by 2, so we can deduce that at least one of the terms
$$|\mathrm{Orb}(A_1)|, |\mathrm{Orb}(A_2)|, \ldots, |\mathrm{Orb}(A_s)|$$
on the right-hand side is not divisible by 2.

We now know that there is an element A of X such that
$$|\mathrm{Orb}(A)|$$
is not divisible by 2.

Applying the Orbit–stabilizer Theorem to A, we have

$$|G| = 20 = |\operatorname{Orb}(A)| \times |\operatorname{Stab}(A)|.$$

Since $|\operatorname{Orb}(A)|$ is not divisible by 2 and $|G|$ is divisible by 4, it follows that $|\operatorname{Stab}(A)|$ is divisible by 4. From this we deduce that

$$|\operatorname{Stab}(A)| \geq 4.$$

So far we have established the existence of a *subgroup*, $\operatorname{Stab}(A)$, of G of order at least 4. If we could show that its order is at most 4, the proof would be complete.

So far we have used the partition equation of the group action and the Orbit–stabilizer Theorem. We have not yet exploited the fact that the elements of X are subsets of the group G. We shall now exploit this fact as we inspect carefully the consequences of the definition of stabilizer.

Suppose that g is any element of $\operatorname{Stab}(A)$. The definition of $\operatorname{Stab}(A)$ means that

$$g \wedge A = A.$$

Now choose a fixed element a in A. Because the action is defined by left multiplication, it follows that

$$ga \in g \wedge A = A.$$

We emphasize: $ga \in A$ for every element g in $\operatorname{Stab}(A)$.

If we consider all the elements of the form ga, for g in $\operatorname{Stab}(A)$, we obtain the *right* coset

$$\operatorname{Stab}(A)a$$

of $\operatorname{Stab}(A)$. We have just shown that all elements of this right coset are in A. Hence

$$\operatorname{Stab}(A)a \subseteq A.$$

Thus, $\operatorname{Stab}(A)a$ cannot have more elements than A and so

$$|\operatorname{Stab}(A)a| \leq |A| = 4.$$

However, right cosets have the same number of elements as the subgroup from which they are formed. Thus

$$|\operatorname{Stab}(A)| = |\operatorname{Stab}(A)a| \leq 4.$$

This provides the other inequality that we wanted and we deduce that

$$|\operatorname{Stab}(A)| = 4.$$

We have shown the existence of a subgroup $\operatorname{Stab}(A)$ of G which has order 4, as required.

Thus, we have shown that any group G of order 20 has a Sylow 2-subgroup of order 4. ♦

Exercise 1.1

Adapt the argument used in Example 1.1 to show that any group of order 20 must have a Sylow 5-subgroup of order 5.
You should define an appropriate group action, but you need not verify the axioms for the action that you have defined.

To complete the arguments used in Example 1.1 and in the solution to Exercise 1.1, we need to verify that the group actions used are actions. We now ask you to do this in the general case.

Exercise 1.2

Let G be a finite group and let X be the set of all subsets of G which have exactly k elements. Define
$$g \wedge A = gA$$
for every element $g \in G$ and $A \in X$.

Show that this defines an action of G on X by proving that, for all $g, h \in G$ and $A \in X$:

(a) $g \wedge A$ is an element of X, i.e. it is a k-element subset of G;

(b) $e \wedge A = A$;

(c) $(gh) \wedge A = g \wedge (h \wedge A)$.

We now examine the key steps in Example 1.1 and in the solution to Exercise 1.1 to see what we shall need to do to generalize the proofs.

The general situation is that we have a group G of order n and a prime p dividing n, with p^α the highest power of p dividing n. Generalizing the key steps in Example 1.1 and Exercise 1.1, we have the following.

(a) Define X to be the set consisting of all subsets of G containing p^α elements.

(b) Define an action of G on X by left multiplication. That is,
$$g \wedge A = gA, \quad g \in G, \, A \in X.$$

The solution to Exercise 1.2 shows that this is a group action.

(c) Show that $|X|$ is *not* divisible by p.

(d) If
$$\mathrm{Orb}(A_1), \mathrm{Orb}(A_2), \ldots, \mathrm{Orb}(A_s)$$
are the distinct orbits for this group action, deduce from the partition equation
$$|X| = |\mathrm{Orb}(A_1)| + |\mathrm{Orb}(A_2)| + \cdots + |\mathrm{Orb}(A_s)|$$
that at least one orbit has order not divisible by p.

(e) If A is an element in X such that $|\mathrm{Orb}(A)|$ is not divisible by p, use the Orbit–stabilizer Theorem
$$|G| = |\mathrm{Orb}(A)| \times |\mathrm{Stab}(A)|$$
to deduce that $|\mathrm{Stab}(A)|$ is divisible by p^α. Hence
$$|\mathrm{Stab}(A)| \geq p^\alpha.$$

(f) Show that, for a fixed element a of A, the right coset
$$\mathrm{Stab}(A)a \subseteq A$$
and, hence, that
$$|\mathrm{Stab}(A)| = |\mathrm{Stab}(A)a| \leq |A| = p^\alpha.$$

This step does not use the fact that we have selected an A for which the size of the orbit is not divisible by p.

(g) Conclude that $\mathrm{Stab}(A)$ has p^α elements and so is a subgroup of G of order p^α.

Since both $\mathrm{Stab}(A)a$ and A have p^α elements and $\mathrm{Stab}(A)a \subseteq A$, we have $\mathrm{Stab}(A)a = A$.

We can extract three useful results from the above proof strategy for Sylow's First Theorem, which we summarize in the following lemma.

Lemma 1.1

Let G be a finite group and let X be the set of all k-element subsets of G. If G acts by left multiplication on X and if $A \in X$, then:

(a) $|\mathrm{Stab}(A)|$ divides $|A|$;

(b) if k is a power of a prime p, then $\mathrm{Stab}(A)$ is a p-group;

(c) if k is the maximal power of p dividing the order of G and if p does not divide $|\mathrm{Orb}(A)|$, then
$$\mathrm{Stab}(A)a = A, \quad \text{for any } a \in A.$$

The proofs of parts (a) and (b) are given after the proof of Sylow's First Theorem, in Section 2. The proof of part (c) follows directly from the marginal note next to step (g) above.

In Example 1.1, we could replace the order, 20, of the group by a general value, n, and the prime, 2, by a general prime, p, throughout, and still have a valid proof *except* at one point. One crucial step in Example 1.1 involved observing that the number of elements of X was not divisible by 2. For a general proof we would have to show the corresponding statement that, if p^α is the highest power of p dividing the order of the group, then the number of p^α-element subsets of the group is not divisible by p. That is, we would have to show that the binomial coefficient

$$^nC_{p^\alpha}$$

is not divisible by p.

We shall consider such divisibility properties of binomial coefficients in Section 2. Once we have shown that

$$^nC_{p^\alpha}$$

is not divisible by p, then the discussion above shows that all the other steps in the argument do generalize and we have a proof of Sylow's First Theorem.

Exercise 1.3

(a) Show that $^{24}C_3$ is not divisible by 3, and hence that any group of order 24 has a subgroup of order 3.

(b) Show that any group of order 24 has a subgroup of order 8.

Exercise 1.4

Let G be a group of order 15. Show that G has a subgroup H of order 3 and a subgroup K of order 5.

The other Sylow theorems, which we discuss later in this unit, are concerned with the *number* of different Sylow p-subgroups corresponding to the prime p. To conclude this section, we investigate an example which will indicate the nature of these results and will lead to a formulation of Sylow's Second Theorem.

Example 1.2

We continue the investigation of groups of order 15, using the same notation as in Exercise 1.4.

Let G be a group of order 15, let X be the set of all 5-element subsets of G and let

$$\text{Orb}(A_1), \ldots, \text{Orb}(A_s)$$

be the distinct orbits under the action of G on X by left multiplication.

Since the Sylow 5-subgroups contain 5 elements, they are elements of X. Thus each 5-subgroup must be contained in some orbit of the action. We shall show that the orbits which contain these subgroups are precisely the ones whose size is not divisible by 5 and that each such orbit contains exactly one Sylow 5-subgroup. Thus, counting the number of Sylow 5-subgroups is the same as counting the number of orbits whose size is not divisible by 5.

This will be accomplished in three steps.

(a) If $A \in X$ is such that 5 does not divide $|\text{Orb}(A)|$, then $\text{Orb}(A)$ contains a Sylow 5-subgroup.

(b) If $\text{Orb}(A)$ contains a Sylow 5-subgroup then 5 does not divide $|\text{Orb}(A)|$.

(c) Any orbit can contain at most one Sylow 5-subgroup.

Let us now go through these steps in detail.

(a) Since 5 does not divide $|\operatorname{Orb}(A)|$, we know from the proof strategy of Sylow's First Theorem that $\operatorname{Stab}(A)$ is a Sylow 5-subgroup. Call this subgroup H.

We have shown, in Lemma 1.1, that for some fixed element a in A,

$$Ha = A.$$

Now the orbit of A includes

$$a^{-1} \wedge A = a^{-1}A = a^{-1}Ha.$$

But $a^{-1}Ha$ is the conjugate of H by a^{-1}, and so, as you could easily verify, is a subgroup of G with the same order as H. In other words, $a^{-1}Ha$ is a Sylow 5-subgroup contained in $\operatorname{Orb}(A)$.

In general, if H is a subgroup of a group G and $g \in G$, then gHg^{-1} is also a subgroup of G of the same order as H.

(b) Suppose that $\operatorname{Orb}(A)$ contains the Sylow 5-subgroup H. This means that $\operatorname{Orb}(H) = \operatorname{Orb}(A)$.

In order to show that the size of this orbit is not divisible by 5, we shall show that $\operatorname{Stab}(H) = H$ and use the Orbit–stabilizer Theorem.

Let $g \in G$. Then

$$g \wedge H = gH.$$

Now $gH = H$ if and only if $g \in H$. That is, $g \in \operatorname{Stab}(H)$ if and only if $g \in H$. Thus $\operatorname{Stab}(H) = H$.

Applying the Orbit–stabilizer Theorem,

$$|G| = 15 = |\operatorname{Orb}(H)| \times |\operatorname{Stab}(H)| = |\operatorname{Orb}(H)| \times 5.$$

This shows that $|\operatorname{Orb}(H)|$, and hence $|\operatorname{Orb}(A)|$, is not divisible by 5.

(c) Suppose $\operatorname{Orb}(A)$ contains the Sylow 5-subgroup H. Then, as before,

$$\operatorname{Orb}(A) = \operatorname{Orb}(H) = \{gH : g \in G\}.$$

Thus the orbit of A consists of all left cosets of H, only one of which is a subgroup, namely H itself.

Only H is a subgroup since only H contains e.

What we have just done shows that there is exactly one Sylow 5-subgroup in each of the orbits whose size is *not* divisible by 5 and no Sylow 5-subgroup in any orbit whose size is divisible by 5. Furthermore, all the Sylow 5-subgroups are accounted for in this way since, for any Sylow 5-subgroup H, we have $H \in X$ and $H \in \operatorname{Orb}(H)$.

We can now use the partition equation for the action to say more about the number of Sylow 5-subgroups.

Suppose that G has m distinct subgroups of order 5, which we call

$$H_1, \ldots, H_m.$$

Each of these belongs to a distinct orbit and so we can write the partition equation as

$$|X| = 3003$$
$$= |\operatorname{Orb}(H_1)| + \cdots + |\operatorname{Orb}(H_m)| + |\operatorname{Orb}(A_{m+1})| + \cdots + |\operatorname{Orb}(A_s)|,$$

where the sizes of the first m orbits are not divisible by 5 and the sizes of the remaining orbits are.

By the Orbit–stabilizer Theorem,
$$|G| = 15 = |\operatorname{Orb}(H_i)| \times |\operatorname{Stab}(H_i)|, \quad i = 1, \ldots, m.$$
We showed above that
$$\operatorname{Stab}(H_i) = H_i, \quad i = 1, \ldots, m$$
and so
$$|\operatorname{Stab}(H_1)| = \cdots = |\operatorname{Stab}(H_m)| = 5.$$
Hence
$$|\operatorname{Orb}(H_1)| = \cdots = |\operatorname{Orb}(H_m)| = 3.$$
The sizes of the remaining orbits are divisible by 5 so we have
$$|X| = 3003 = 3m + (\text{a multiple of } 5).$$
Reducing this equation modulo 5 gives
$$3 \equiv 3m \pmod{5}.$$
Multiplying both sides by 2, and observing that $2 \times 3 = 6$ is congruent to 1 modulo 5, gives
$$m \equiv 1 \pmod{5}.$$
Thus, in any group of order 15, the number of Sylow 5-subgroups is congruent to 1 modulo 5. ♦

In Section 2 we shall show that the argument above generalizes to give the following theorem.

Theorem 1.2 Sylow's second theorem

Let G be a finite group of order n and let p be a prime dividing n. Then the number of distinct Sylow p-subgroups of G is congruent to 1 modulo p.

We ask you to take Sylow's Second Theorem on trust for now. To indicate its use, we now ask you to apply it to classify all groups of order 15.

Exercise 1.5

Let G be a group of order 15.

(a) Show that two distinct Sylow 5-subgroups of G must have trivial intersection.

(b) Assume that G has m Sylow 5-subgroups and that Sylow's Second Theorem holds, i.e. that m is congruent to 1 modulo 5.
By counting elements of order 5, show that G has exactly one subgroup of order 5 and hence only 4 elements of order 5.

(c) Let l be the number of Sylow 3-subgroups of G.
Determine, in terms of l, the number of elements of G order 3.

(d) Assume that Sylow's Second Theorem holds, i.e. that l is congruent to 1 modulo 3.
By counting the number of elements of G of orders 1, 3 and 5, show (by contradiction) that G has an element of order 15 and, hence, that G is cyclic.

So, using Sylow's First and Second Theorems, we have been able to show that the only group of order 15 is $\mathbb{Z}_{15}$.

2 SYLOW'S FIRST AND SECOND THEOREMS

Our main aim in this section is to give formal proofs of Sylow's First and Second Theorems.

We restate the First Theorem here for reference.

> *Sylow's first theorem*
>
> Let G be a finite group of order n and let p be a prime dividing n, where p^α is the highest power of p dividing n.
> Then G has a subgroup of order p^α.

The theorem prompted us to make the following definition.

> *Definition 2.1 Sylow p-subgroup*
>
> Let G be a finite group of order n and let p be a prime dividing n.
> Let p^α be the highest power of p dividing n.
> Then a subgroup of G of order p^α is called a **Sylow p-subgroup** of G.

As the discussion in Section 1 indicated, the proof of Sylow's First Theorem will use the action of G defined by left multiplication on the set X of all p^α-element subsets of G.

In Section 1, we noted that a crucial step in the argument was to show that the size of X is not divisible by p. Since the size of X is given by the number of ways of choosing p^α elements from n elements, we have

$$|X| = {}^nC_{p^\alpha}.$$

We therefore need to consider which powers of p can divide this binomial coefficient.

Example 1.1 from Section 1 and the note next to the solution to Exercise 1.3 give some clues as to how a general proof can be constructed. In Example 1.1 we had a group of order 20 and considered the set X of all 4-element subsets of the group. Writing out the calculation of this binomial coefficient in detail, we have

$$\begin{aligned}|X| &= {}^{20}C_4 \\ &= \frac{20 \times (20-1) \times (20-2) \times (20-3)}{4 \times (4-1) \times (4-2) \times (4-3)} \\ &= \frac{5 \times (20-1) \times (10-1) \times (20-3)}{1 \times (4-1) \times (2-1) \times (4-3)}.\end{aligned}$$

The important observation, for present purposes, is that any power of 2 which divides a term in the numerator also divides the *corresponding* term in the denominator, and conversely.

This example suggests that we write the general formula for the binomial coefficient that we require in the form

$$\begin{aligned}{}^nC_{p^\alpha} &= \frac{n \times (n-1) \times \cdots \times (n-(p^\alpha - 1))}{p^\alpha \times (p^\alpha - 1) \times \cdots \times 1} \\ &= \frac{(n-0) \times (n-1) \times \cdots \times (n-(p^\alpha - 1))}{(p^\alpha - 0) \times (p^\alpha - 1) \times \cdots \times (p^\alpha - (p^\alpha - 1))}.\end{aligned}$$

Given the conditions of Sylow's First Theorem, we can write
$$n = p^\alpha z,$$
where z is not divisible by p.

The expression for the binomial coefficient becomes
$$^{p^\alpha z}C_{p^\alpha} = \frac{(p^\alpha z - 0) \times (p^\alpha z - 1) \times \cdots \times (p^\alpha z - (p^\alpha - 1))}{(p^\alpha - 0) \times (p^\alpha - 1) \times \cdots \times (p^\alpha - (p^\alpha - 1))}.$$

Each term in the numerator is of the form
$$p^\alpha z - k,$$
where $k = 0, \ldots, p^\alpha - 1$.

The corresponding terms in the denominator are of the form
$$p^\alpha - k,$$
where $k = 0, \ldots, p^\alpha - 1$.

The case of $^{20}C_4$ considered above suggests that the highest powers of p dividing
$$p^\alpha z - k$$
and
$$p^\alpha - k$$
are the same, for each value of k from 0 to $p^\alpha - 1$ (and are in fact the highest power of p dividing k). We ask you to establish this result in the next exercise, which uses the notation described above.

Exercise 2.1

(a) Let k satisfy
$$0 < k < p^\alpha.$$

  (i) Show that, if p^β divides $p^\alpha z - k$, then $\beta < \alpha$.

  (ii) Hence show that p^β divides $p^\alpha z - k$ if and only if p^β divides k.

(b) Deduce that, if p does not divide z, then
$$^{p^\alpha z}C_{p^\alpha}$$
is not divisible by p.

As we indicated in Section 1, the fact that if p is a prime that does not divide z then
$$^{p^\alpha z}C_{p^\alpha}$$
is not divisible by p is the key to providing a general proof of Sylow's First Theorem.

The special cases dealt with in Section 1 can now be translated directly into a general proof, which we present here for reference.

Proof of Sylow's First Theorem

Let X be the set of all subsets of G having p^α elements.

Let G act on X by left multiplication. Suppose that the distinct orbits under this action are
$$\operatorname{Orb}(A_1), \ldots, \operatorname{Orb}(A_s).$$

By the partition equation we have
$$|X| = |\operatorname{Orb}(A_1)| + \cdots + |\operatorname{Orb}(A_s)|.$$

By the result of Exercise 2.1, the left-hand side is not divisible by p. Hence at least one term on the right-hand side is not divisible by p.

You have shown in Exercise 1.2 that this is a group action.

The use of group actions to prove Sylow's First Theorem is due to H. Wielandt and is relatively recent (1959), although the result itself has been established for over a century.

Let A be an element of X such that $|\operatorname{Orb}(A)|$ is not divisible by p. By the Orbit–stabilizer Theorem,

$$|G| = |\operatorname{Orb}(A)| \times |\operatorname{Stab}(A)|.$$

But $|G|$ is divisible by p^α and $|\operatorname{Orb}(A)|$ is not divisible by p. It follows that $|\operatorname{Stab}(A)|$ is divisible by p^α. Thus

$$|\operatorname{Stab}(A)| \geq p^\alpha.$$

Now consider any element g of $\operatorname{Stab}(A)$ and a fixed element a of A. Since $\operatorname{Stab}(A)$ is the stabilizer of A,

$$ga \in g \wedge A = A.$$

Thus, $ga \in A$ for every element $g \in \operatorname{Stab}(A)$. This implies that

$$\operatorname{Stab}(A)a \subseteq A.$$

> This particular part of the proof does *not* use the fact that $|\operatorname{Orb}(A)|$ is not divisible by p.

Hence

$$|\operatorname{Stab}(A)| = |\operatorname{Stab}(A)a| \leq |A| = p^\alpha.$$

It follows that $|\operatorname{Stab}(A)| = p^\alpha$. Thus $\operatorname{Stab}(A)$ is a subgroup of G with order p^α, completing the proof. ∎

The action of G on the set X of p^α-element subsets of G provides rather more information than is actually used in the proof of Sylow's First Theorem.

In the proof, the existence of a Sylow p-subgroup was established by showing that the stabilizer of some element $A \in X$ had the correct order. The proof shows that the stabilizer of *any* $A \in X$, with $|\operatorname{Orb}(A)|$ not divisible by p, will be a Sylow p-subgroup of G. We shall show later that all Sylow p-subgroups arise in this way.

Some parts of the proof did not depend on $|\operatorname{Orb}(A)|$ not being divisible by p.

In particular, if $a \in A$ and $g \in \operatorname{Stab}(A)$, then

$$ga \in g \wedge A = A,$$

> This formalizes the proof of Lemma 1.1(a).

and so all the elements of the right coset $\operatorname{Stab}(A)a$ are contained in A. On the other hand, if a is any element of A, then a is in the right coset $\operatorname{Stab}(A)a$.

Thus any $A \in X$ is the union of some right cosets of its stabilizer $\operatorname{Stab}(A)$, those given by elements of A.

Since distinct right cosets are disjoint, this means that A consists of a union of disjoint right cosets of $\operatorname{Stab}(A)$.

Because all the right cosets have the same size as $\operatorname{Stab}(A)$,

$$|\operatorname{Stab}(A)| \text{ divides } |A|,$$

for any A in X.

Now, as A has p^α elements, we can deduce that the group $\operatorname{Stab}(A)$ is a p-group, for any element $A \in X$, whether or not $\operatorname{Stab}(A)$ is a Sylow p-subgroup.

> This formalizes the proof of Lemma 1.1(b).

Let us now move on towards proving Sylow's Second Theorem. We begin with a preliminary result.

Let G and X be as defined in Sylow's First Theorem. As earlier, we write

$$|G| = n = p^\alpha z,$$

where z is not divisible by p. Then, for any element $A \in X$, from the Orbit–stabilizer Theorem we have

$$|G| = p^\alpha z = |\operatorname{Orb}(A)| \times |\operatorname{Stab}(A)|.$$

Since, as we have just seen, the stabilizer is a p-group, it follows that $|\operatorname{Orb}(A)|$ is divisible by z.

This result is sufficiently useful to state it as a lemma.

> **Lemma 2.1**
>
> Let G be a finite group of order $p^\alpha z$, where p is a prime not dividing z. Let X be the set of p^α-element subsets of G. Then, for the action of G on X by left multiplication, the size of every orbit is divisible by z.

We can use this lemma to obtain useful information about the *number* of Sylow p-subgroups, as we shall do in the proof below.

In Section 1, we considered a special case and found that the number of Sylow p-subgroups was congruent to 1 modulo p. This generalized to give the Sylow's Second Theorem, which we restate here for reference.

> **Sylow's second theorem**
>
> Let G be a finite group of order n and let p be a prime dividing n. Then the number of distinct Sylow p-subgroups of G is congruent to 1 modulo p.

Proof

Our proof uses the same group action and notation as the proof of Sylow's First Theorem.

The strategy for the proof is to show that Sylow p-subgroups are exactly the stabilizers of elements whose orbits have size not divisible by p and that each such orbit contains exactly one Sylow p-subgroup. We then use the partition equation

$$|X| = |\operatorname{Orb}(A_1)| + \cdots + |\operatorname{Orb}(A_s)|$$

to obtain the required result.

This strategy generalizes a result we obtained in Example 1.2 of Section 1.

We start by showing that, if $A \in X$ is such that $|\operatorname{Orb}(A)|$ is not divisible by p, then $\operatorname{Orb}(A)$ contains a Sylow p-subgroup.
Under these conditions we know, as a consequence of our proof of Sylow's First Theorem, that $H = \operatorname{Stab}(A)$ is a Sylow p-subgroup. We also know, by Lemma 1.1, that, for a fixed element $a \in A$,

$$Ha = A.$$

Now the orbit of A includes

$$a^{-1} \wedge A = a^{-1}A = a^{-1}Ha.$$

But $a^{-1}Ha$ is the conjugate of H by a^{-1}, and so is a subgroup of G with the same order as H. In other words, $a^{-1}Ha$ is a Sylow p-subgroup contained in $\operatorname{Orb}(A)$.

Thus every orbit whose size is not divisible by p contains as an element a Sylow p-subgroup.

Conversely, consider a Sylow p-subgroup H of G. By the definition of X, the subgroup H is an element of X because $|H| = p^\alpha$.
We shall show that the stabilizer of H is H itself, that is

$$\operatorname{Stab}(H) = H,$$

and hence that H is in an orbit, $\operatorname{Orb}(H)$, whose size is not divisible by p. By the definition of the action, if $g \in G$ then

$$g \wedge H = gH,$$

a left coset of H in G. However we know that $gH = H$ if and only if $g \in H$. So $g \wedge H = H$, i.e. $g \in \operatorname{Stab}(H)$, if and only if $g \in H$. This shows that $\operatorname{Stab}(H) = H$.

The Orbit–stabilizer Theorem gives

$$p^\alpha z = |G|$$
$$= |\operatorname{Orb}(H)| \times |\operatorname{Stab}(H)|$$
$$= |\operatorname{Orb}(H)| \times |H|$$
$$= |\operatorname{Orb}(H)| \times p^\alpha.$$

Hence, $|\operatorname{Orb}(H)| = z$ and is not divisible by p, and of course $H \in \operatorname{Orb}(H)$.

We now know that an orbit contains a Sylow p-subgroup if and only if the size of the orbit is not divisible by p. The last main step is to show that an orbit can contain at most one Sylow p-subgroup.

Let H be a Sylow p-subgroup. Then $\operatorname{Orb}(H)$ consists of all

$$g \wedge H = gH, \quad g \in G,$$

that is, all right cosets of H. However, the only right coset of the subgroup H which is a subgroup of G is H itself. Hence no orbit can contain more than one Sylow p-subgroup.

Thus we have established that each orbit whose size is not divisible by p contains exactly one Sylow p-subgroup and the other orbits whose size is divisible by p contain no Sylow p-subgroups. Hence the number of Sylow p-subgroups is the number of orbits whose size is not divisible by p.

We now separate the terms on the right-hand side of the partition equation

$$|X| = |\operatorname{Orb}(A_1)| + \cdots + |\operatorname{Orb}(A_s)|$$

into those which are divisible by p and those which are not.

Suppose that G has m distinct Sylow p-subgroups, and so the action has m distinct orbits whose size is not divisible by p. We write the partition equation as

$$|X| = |\operatorname{Orb}(A_1)| + \cdots + |\operatorname{Orb}(A_m)| + |\operatorname{Orb}(A_{m+1})| + \cdots + |\operatorname{Orb}(A_s)|,$$

where the first m terms on the right are not divisible by p and the remaining terms are.

For each of the first m terms, $\operatorname{Stab}(A_i)$ is a Sylow p-subgroup and the Orbit–stabilizer Theorem gives

$$|G| = p^\alpha z = |\operatorname{Orb}(A_i)| \times |\operatorname{Stab}(A_i)| = |\operatorname{Orb}(A_i)| \times p^\alpha, \quad i = 1, \ldots, m.$$

Thus

$$|\operatorname{Orb}(A_i)| = z, \quad i = 1, \ldots, m.$$

By Lemma 2.1, *all* the terms in the partition equation are divisible by z. As we have seen, the first m are z. The remaining terms are divisible both by z and by p and, because p and z are coprime, are divisible by pz.

Hence,

$$|X| = mz + (\text{terms divisible by } pz),$$

which we can write as

$$|X| = mz + kpz,$$

for some integer k.

Before continuing, let us paraphrase this result that we have just obtained. Let G be *any* group of order $n = p^\alpha z$, where p is a prime which does not divide z. Let m be the number of Sylow p-subgroups of G. Then there exists an integer k such that

$$|X| = {}^nC_{p^\alpha} = mz + kpz.$$

The integer k will depend on the specific group G.

We now apply this to the special case of the cyclic group $\mathbb{Z}_n$ of order n. In this case we know that there is exactly one subgroup belonging to each divisor of n. In particular, there is exactly one subgroup of order p^α, and so $m = 1$ for $\mathbb{Z}_n$.

Hence, there exists an integer k', such that

$$^nC_{p^\alpha} = z + k'pz.$$

We now have two expressions for the binomial coefficient, which we equate:

$$mz + kpz = z + k'pz.$$

Dividing through by z, we have

$$m + kp = 1 + k'p.$$

Hence, $m - 1 = k'p - kp = p(k' - k)$. So p divides $m - 1$ and

$$m \equiv 1 \pmod{p}.$$

This completes the proof. ∎

Sylow's Second Theorem places sufficient restriction on the number of distinct Sylow p-subgroups for us to be able to make deductions about groups of some specific orders.

Exercise 2.2

Let G be a group of order 91.

(a) Show that G has only one Sylow 13-subgroup.

(b) By investigating the number of elements of order 7, show (by contradiction) that G is cyclic.

We have discussed two examples of groups of order pq, where p and q are distinct primes, namely 15 and 91. On the basis of these it would be tempting to believe that all such groups are cyclic. The existence of the dihedral group D_3 of order $6 = 2 \times 3$ is enough to dispel this belief.

However, it is possible to make, and prove, a modified version of this conjecture. The next exercise provides a first step in this direction.

Exercise 2.3

(a) Let G be a group of order pq, where p and q are primes with $p < q$. Show that G has exactly one Sylow q-subgroup.

(b) By considering the order of the conjugates of the Sylow q-subgroup of G, show that it is normal in G.

Exercise 2.3 provides a complete generalization of *part* of what we discovered about our earlier examples. For groups of this type, the Sylow q-subgroup (the one belonging to the larger prime) is unique and hence normal.

We observe that the proof of the normality of the Sylow q-subgroup in Exercise 2.3 did not use the fact that we were dealing with a Sylow subgroup, but only that there was a unique subgroup of that order. The argument in the solution therefore provides a proof of the following lemma.

Lemma 2.2

If G is a group having only one subgroup of a particular order, then this subgroup is normal.

We have now established two of the three results that are collectively known as Sylow's Theorems. The final result also concerns the number of Sylow p-subgroups. We have already shown that this number is congruent to 1 modulo p. It must also be a divisor of the order of the group, although establishing this requires a different group action from the proofs above. We deal with this matter in Section 3.

Finally, as we observed in Corollary 1.1, the Sylow p-subgroups are p-groups and, therefore, contain subgroups corresponding to each divisor of their order. Thus, if G has a Sylow p-subgroup of order p^α, then G also has subgroups of orders

$$p, p^2, \ldots, p^{\alpha-1}.$$

We shall consider the number of such subgroups, and provide an alternative proof of their existence, in Section 5.

3 SYLOW'S THIRD THEOREM (AUDIO-TAPE SECTION)

In the process of classifying groups of a particular order, knowledge of subgroups is sometimes helpful. The Sylow theorems provide information about subgroups, based only on the order of the group. In particular, as we have seen in some examples, for certain orders the groups must possess a *normal* Sylow p-subgroup. This knowledge can be useful in reducing the number of possibilities to be considered.

We have established two results about a group G of order $n = p^\alpha z$, where p does not divide z. Firstly, G has at least one subgroup of order p^α. Secondly, the number, m say, of such subgroups is congruent to 1 modulo p.

In this section we shall establish that m divides the order of G. This restriction on m often yields useful information.

In fact, we shall establish rather more than this numerical result. We already know that the conjugate of any Sylow p-subgroup is another Sylow p-subgroup. We shall prove the converse: that any two Sylow p-subgroups *must* be conjugate.

> In fact we know that the conjugate of *any* subgroup is a subgroup of the same order.

We shall use one group action to establish that Sylow p-subgroups are conjugate and another, based on this conjugacy, to derive the result that m divides $|G|$.

You should now listen to the audio programme for this unit, referring to the tape frames below when asked to during the programme.

1

Aim

To show that:

the number of Sylow p-subgroups of a group G must divide the order of the group

Notation

Group G $|G| = p^\alpha z$, p prime, $p \nmid z$

Sylow p-subgroups of G:
$$H_1, \ldots, H_m$$

Want

m divides $|G|$

2

Strategy

- Show any two of
$$H_1, \ldots, H_m$$
are conjugate

- Deduce m divides $|G|$ (standard counting arguments)

Method

Group actions: Orbit–stabilizer Theorem
                partition equation

3

First step

Let H and K be any two Sylow p-subgroups

Aim to show H and K are conjugate

Define action

Group acting: H

Set X: all left cosets of K in G

Action: $h \wedge (gK) = (hg)K$, $h \in H$, $gK \in X$

Is this a group action?

Exercise 3.1

Show:

(a) $e \wedge (gK) = gK$ for all $gK \in X$

(b) $h_1 \wedge (h_2 \wedge (gK)) = (h_1 h_2) \wedge (gK)$
      for all $h_1, h_2 \in K$ and all $gK \in X$

3A

Solution 3.1

(a) $e \wedge (gK) = (eg)K = gK$

(b) $h_1 \wedge (h_2 \wedge (gK)) = h_1 \wedge ((h_2 g)K)$
$= (h_1(h_2 g))K$
$= ((h_1 h_2)g)K$
$= (h_1 h_2) \wedge (gK)$

left multiplication always gives a group action

4

Size of X

K has p^α elements

G has $p^\alpha z$ elements *(z is index of K in G)*

K has z cosets in G

$|X| = z$

5

Consequences of Orbit–stabilizer Theorem

If $gK \in X$,

$|H| = |\text{Orb}(gK)| \times |\text{Stab}(gK)|$
↑
p^α

$|\text{Orb}(gK)|$ divides p^α

$|\text{Orb}(gK)|$ is p^β $(0 \leq \beta \leq \alpha)$

6

Partition equation

If there are s different orbits,

$|X| = p^{\beta_1} + \cdots + p^{\beta_s}$ *(p^{β_i} is size of ith orbit)*
↑
z

z is *not* divisible by p

So not all terms on right can be divisible by p, i.e. at least one $p^{\beta_i} = 1$

So, for *some* coset $aK \in X$,

$|\text{Orb}(aK)| = 1$

7

H and K are conjugate

For all $h \in H$:

$h \wedge (aK) = aK$

$(ha)K = aK$ — *coset equality*

$a^{-1}ha \in K$

$h \in aKa^{-1}$

$H \subseteq aKa^{-1}$

$H = aKa^{-1}$ — $|H| = |K|$, $|aKa^{-1}| = |K|$ since conjugate subgroups have same order

8

Progress

Completed: any two Sylow p-subgroups are conjugate

To do: m divides $|G|$

9

A new action

Group acting: G

Set X: $\{H_1, \ldots, H_m\}$ — set of Sylow p-subgroups of G

Action: $g \wedge H_i = gH_ig^{-1}$ — conjugation; another Sylow p-subgroup

Is this a group action?

Exercise 3.2

Show:

(a) $e \wedge H_i = H_i, \quad i = 1, \ldots, m$

(b) $g_1 \wedge (g_2 \wedge H_i) = (g_1 g_2) \wedge H_i$
    for all $g_1, g_2 \in G$ and $i = 1, \ldots, m$

9A

Solution 3.2

(a) $e \wedge H_i = eH_ie^{-1} = eH_ie = H_i$

(b) $g_1 \wedge (g_2 \wedge H_i) = g_1 \wedge (g_2 H_i g_2^{-1})$

$= g_1(g_2 H_i g_2^{-1})g_1^{-1}$

$= (g_1 g_2) H_i (g_2^{-1} g_1^{-1})$

$= (g_1 g_2) H_i (g_1 g_2)^{-1}$

$= (g_1 g_2) \wedge H_i$

10

The new action

Know

There is only one orbit,

namely $\text{Orb}(H_1) = \text{Orb}(H_i)$, $i = 2, \ldots, m$

Hence $|\text{Orb}(H_1)| = m$

> all Sylow p-subgroups are conjugate

Orbit–stabilizer Theorem

$|G| = |\text{Orb}(H_1)| \times |\text{Stab}(H_1)|$

$|G| = m \times |\text{Stab}(H_1)|$

So m divides $|G|$

11

Sylow's Third Theorem

All Sylow p-subgroups of a group G are conjugate

The number of Sylow p-subgroups divides the order of G

12

Corollary

If $|G| = p^\alpha z$ and m is the number of Sylow p-subgroups

then m divides z

> usual notation

> by Sylow's Second Theorem, $m \equiv 1 \pmod{p}$

For reference, we present a complete statement of the Sylow theorems.

> **Theorem 3.1 The Sylow theorems**
>
> Let G be a finite group of order $n = p^\alpha z$ where p is a prime not dividing z. Let m be the number of subgroups of G of order p^α. Then:
>
> (a) G has a Sylow p-subgroup, that is $m \geq 1$;
>
> (b) $m \equiv 1 \pmod{p}$, that is m is of the form $1 + kp$ for some integer k;
>
> (c) all the Sylow p-subgroups of G are conjugate;
>
> (d) m divides $|G|$.

Because m is congruent to 1 modulo p, it follows that p^α and m are coprime. Thus m divides $n = p^\alpha z$ implies that m divides z. This form of the fourth result is slightly easier to use for large values of n, and so we state it as a corollary.

> **Corollary 3.1**
>
> Let G be a finite group of order $n = p^\alpha z$ where p is a prime not dividing z. Let m be the number of p-Sylow subgroups of G. Then m divides z.

The fact that the number of Sylow p-subgroups divides the order of the group provides shorter proofs of some of the results that have already been obtained.

Exercise 3.3

Let G be a group of order 15. Suppose that G has l Sylow 3-subgroups and m Sylow 5-subgroups.

(a) Show that $l = 1$.

(b) Show that $m = 1$.

(c) Show that G is cyclic.

Exercise 3.4

Let G be a group of order 6. Show that G has a unique Sylow 3-subgroup but could have either one or three Sylow 2-subgroups.

By considering the groups of order 6, show that both possibilities occur.

4 APPLICATIONS OF THE SYLOW THEOREMS

Having proved the Sylow theorems, we now investigate some applications. The section consists mainly of exercises of three basic types.

Firstly, we ask you to find the number and type of Sylow subgroups of some specific groups. Secondly, we ask you to use the Sylow theorems to classify groups of a particular order. Thirdly, we ask you to use the Sylow theorems to show that certain groups contain no non-trivial, proper, normal subgroups.

The process of finding the number and type of Sylow subgroups of a group is sometimes referred to as 'finding the Sylow structure of a group'.

This third type of activity needs some comment. When investigating the structure of a group G, the knowledge that G has a non-trivial, proper, normal subgroup, N say, can be useful. We can look at the subgroup N and the quotient group G/N. Both of these groups will be smaller than G and may provide information about the structure of G itself. Thus, a group without such normal subgroups is, in some sense, indecomposable. A group without non-trivial, proper, normal subgroups is called a **simple group**. (The term is somewhat misleading as simple groups may have very complicated structures!) If a group has a unique Sylow p-subgroup for some prime p, then by Lemma 2.2 the subgroup is normal and so, provided the group is not a p-group, the group cannot be simple. We shall mention simple groups again in *Unit GR6*. For now we simply note the term 'simple', and that the Sylow theorems can provide a method of showing that a group is not simple.

There is a loose analogy with prime numbers here.

In a p-group G, the unique Sylow p-subgroup is the whole of G, and so is not a proper subgroup.

The classification theme will also be developed further in *Unit GR6*, where we shall classify all groups of order 12, a task that turns out to be more substantial than the relatively small order might suggest.

We begin here, however, with the first type of activity by considering the Sylow subgroups of some familiar groups.

Exercise 4.1

For each of the following groups, find the Sylow p-subgroups for each prime dividing the order.

(a) $\mathbb{Z}_6$

(b) $\mathbb{Z}_3 \times \mathbb{Z}_{15}$

(c) S_3

We next look at the Sylow subgroups of some dihedral groups. When thinking about such groups, it is often useful to have in mind both their algebraic and geometric forms.

For example, D_7 is, geometrically, the group of symmetries of a regular 7-gon. Its elements are 7 rotations about the centre and 7 reflections in the axes of symmetry. If we denote the rotation through $2\pi/7$ by a and a selected reflection by b, then we obtain the presentation

$$D_7 = \langle a, b : a^7 = e, b^2 = e, ba = a^6 b (= a^{-1} b) \rangle.$$

The rotations are through angles of $0, 2\pi/7, \ldots, 12\pi/7$.

In doing calculations within a dihedral group, sometimes it is easier to work with the presentation, sometimes with the geometry.

As with the separation of permutations into even and odd, we can separate the symmetries in D_7 into direct and indirect symmetries. Less formally, imagine the 7-gon coloured red on one side, blue on the other. Rotations preserve colour, reflections (carried out by turning over) change the colour.

You have met direct and indirect symmetries in the Geometry stream of the course.

This observation makes it clear that conjugates of rotations are rotations. For if r is a rotation and g is any element of the group, consider grg^{-1}. Whatever g does to the colour, i.e. preserves or changes it, g^{-1} does the same. Thus grg^{-1} must preserve colour and is, therefore, a rotation.

Exercise 4.2

Consider the dihedral group D_7.

(a) Show that D_7 has a unique Sylow 7-subgroup.

(b) Using the fact that D_7 is *not* Abelian, show that D_7 has exactly 7 Sylow 2-subgroups.

(c) Deduce that all reflections are conjugate in D_7.

(d) Working from the presentation described above, show that each non-identity rotation is conjugate to itself and its inverse only.

(e) Write down the class equation for D_7 and hence find the centre of the group.

Hint You may find the relation $ba^i = a^{-i}b$, derived from $ba = a^{-1}b$, useful (*Unit IB3*, Theorem 1.1).

Exercise 4.3

Show that there are only two groups of order 14, the cyclic group $\mathbb{Z}_{14}$ and the dihedral group D_7.

Hint Use the Sylow theorems to deduce the existence of an element a of order 7 and an element b of order 2. By considering cosets, show that these two elements generate the group. By considering the conjugate of a by b, deduce the possible relations between a and b.

Many, but not all, of the results that were obtained in the solution to Exercise 4.2 apply to other dihedral groups.

Exercise 4.4

Consider the dihedral group D_{14} of order 28 with presentation

$$D_{14} = \langle a, b : a^{14} = e,\ b^2 = e,\ ba = a^{13}b\ (= a^{-1}b) \rangle.$$

(a) Show that D_{14} has a unique Sylow 7-subgroup.

(b) Show that the elements of D_{14} are of orders 1, 2, 7 and 14 only. Show also that there is exactly one rotation of order 2.

(c) Show that the product of two reflections is a rotation. Deduce that each Sylow 2-Subgroup contains two distinct reflections and the unique rotation of order 2 (as well as the identity).

(d) Show that the subgroup $\{e, a^7\}$ is normal and that it is contained in every Sylow 2-subgroup of D_{14}.

(e) Show that D_{14} has exactly 7 Sylow 2-subgroups.

In Sections 2 and 3 we saw an example of a group of order pq, where p and q are distinct primes, which has a unique (normal) Sylow p-subgroup and a unique (normal) Sylow q-subgroup, from which facts we were able to deduce that the group is cyclic.

The example group had order 15.

However, from our point of view, it is not the fact that the group is *cyclic* that is of particular interest, rather that the group is, in fact, the internal direct product of its unique Sylow subgroups. To show that this is true we need a special case of the Internal Direct Product Theorem from *Unit GR2*.

We begin by reminding you of the statement of the Internal Direct Product Theorem.

Internal direct product theorem

If H_1 and H_2 are subgroups of a group G, then

$$\phi : H_1 \times H_2 \to G$$
$$(h_1, h_2) \mapsto h_1 h_2$$

is an isomorphism if and only if all three of the following conditions hold:

(a) $G = H_1 H_2$;
(b) $H_1 \cap H_2 = \{e\}$;
(c) H_1 and H_2 are normal subgroups of G.

In the case of a *finite* group, we have the following special case.

Theorem 4.1 Internal direct product for finite groups

If H_1 and H_2 are subgroups of a finite group G, then

$$\phi : H_1 \times H_2 \to G$$
$$(h_1, h_2) \mapsto h_1 h_2$$

is an isomorphism if all three of the following conditions hold:

(a) $|G| = |H_1| \times |H_2|$;
(b) $|H_1|$ and $|H_2|$ are coprime;
(c) H_1 and H_2 are normal subgroups of G.

Proof

We are given the third condition of the Internal Direct Product Theorem. It remains to show that $H_1 \cap H_2 = \{e\}$ and that $G = H_1 H_2$.

Now, $H_1 \cap H_2$ is a subgroup of both H_1 and H_2. By Lagrange's Theorem, the order of $H_1 \cap H_2$ must divide both $|H_1|$ and $|H_2|$. Since these two orders are coprime,

$$|H_1 \cap H_2| = 1$$

and so the intersection is trivial, that is $H_1 \cap H_2 = \{e\}$.

Because

$$|G| = |H_1| \times |H_2|,$$

it *appears* that G and $H_1 H_2$ have the same number of elements and are, thus, equal. The only problem that could arise is if two of the products in $H_1 H_2$ are equal. Suppose, therefore, that

$$h_1 h_2 = k_1 k_2$$

where $h_1, k_1 \in H_1$ and $h_2, k_2 \in H_2$. Then, multiplying on the left by k_1^{-1} and on the right by h_2^{-1}, we have

$$k_1^{-1} h_1 = k_2 h_2^{-1}.$$

Since H_1 and H_2 are subgroups, the left-hand side is in H_1 and the right-hand side is in H_2. So both are in $H_1 \cap H_2 = \{e\}$. But

$$k_1^{-1} h_1 = e \quad \Rightarrow \quad h_1 = k_1$$

and, similarly, $h_2 = k_2$.

Thus all the products in $H_1 H_2$ are distinct and, by counting the elements as above,

$$G = H_1 H_2.$$

All the conditions of the Internal Direct Product Theorem are met and so

$$G \cong H_1 \times H_2. \qquad \blacksquare$$

We make two observations about this theorem and proof.

Firstly, the proof can be extended to cover more than two normal subgroups with orders coprime in pairs, the product of whose orders is the order of the group.

Formally, this extension requires the Principle of Mathematical Induction.

Secondly, we shall often apply Theorem 4.1 when the subgroups are unique Sylow subgroups for various primes. In this case the coprimeness of the orders is guaranteed, as is the product property, and normality is a consequence of the uniqueness.

The group of order 15 that we have already looked at in various ways is of this type. We know that it has a unique Sylow 3-subgroup and a unique Sylow 5-subgroup, both of which are cyclic and which have coprime orders. The group is, therefore, the internal direct product of these subgroups. The cyclicity of the group follows from what we know about the direct product of cyclic groups of coprime orders.

Exercise 4.5

(a) Let G be a group of order 221. Show that G is cyclic.

(b) Why does the argument used in the case $|G| = 221$ fail for a group of order 111?

Finally, we ask you to show that groups of particular orders cannot be simple.

Exercise 4.6

Let G be a group of order 110. Show that G has a non-trivial, proper, normal subgroup.

Exercise 4.7

Show that a group of order 132 cannot be simple.

Hint Consider the possibility that none of the Sylow subgroups is unique and show that this leads to a contradiction.

We shall develop the arguments that we have used here in *Unit GR6*, where we shall extend the type of analysis used in the exercise on groups of order 132 to classify groups of order 12.

5 SUBGROUPS OF PRIME POWER ORDER

The Sylow theorems deal with the existence and number of Sylow p-subgroups for any prime p dividing the order of the group. Thus, if G is a group with

$$|G| = p^\alpha z,$$

where p is prime and z is not divisible by p, then the theorems only give information about the subgroups of order p^α, and not about those of order p^β where $\beta < \alpha$.

As we observed earlier, in Section 1, because any Sylow p-subgroup H of a group G is a p-group, then our work on p-groups in *Unit GR4* shows that H has subgroups of each power of p from p^0 to p^α. This argument shows that G has subgroups corresponding to *every* power of p dividing $|G|$.

This was the result we labelled Corollary 1.1.

What is not obvious from this argument is that not only is the number of Sylow p-subgroups congruent to 1 modulo p, but the number of subgroups belonging to a particular integer power of p, say p^β, where $0 \leq \beta \leq \alpha$, is also congruent to 1 modulo p.

The proofs used for the first two of the Sylow theorems can be adapted to deal with powers less than the highest. Such adaptation will give a second proof of the existence of such subgroups and also prove the 'congruent to 1 modulo p' result.

What cannot be generalized, because it is not true in general, is the conjugacy property of Sylow p-subgroups. Indeed, a group may have subgroups of the same prime power order which are not even isomorphic, so certainly cannot be conjugate. This is because conjugate subgroups are always isomorphic since, if H is a subgroup of a group G, then

$$x \mapsto gxg^{-1}, \quad g \in G, x \in H,$$

defines an isomorphism from H to the conjugate subgroup gHg^{-1}.

The next exercise provides an example of a group with non-isomorphic p-subgroups of the same order.

Exercise 5.1

Let D_4 be the dihedral group of order 8 with presentation

$$D_4 = \langle a, b : a^4 = e, b^2 = e, ba = a^{-1}b \, (= a^3 b) \rangle.$$

(a) Show that D_4 has a cyclic subgroup of order 4.
(b) Show that $\{e, a^2, b, a^2 b\}$ is a subgroup of order 4 isomorphic to the Klein group.

The solution to the previous exercise shows that D_4 has two non-isomorphic 2-subgroups of order 4. Thus the subgroups of D_4 with order 4 certainly cannot all be conjugate.

Along with conjugacy, we also have to discard the result that the number of subgroups belonging to a particular prime power is a divisor of the order of the group. As our proof of the divisibility property used conjugacy, it is not entirely surprising that divisibility is lost too. You might wish to verify that D_4 has three subgroups of order 4, one cyclic and two isomorphic to the Klein group, and of course 3 does not divide $8 = |D_4|$.

Our aim, then, is to prove the following theorem.

> **Theorem 5.1 Prime power subgroups**
>
> Let G be a finite group and p be a prime.
> If p^β divides $|G|$ then G has a subgroup of order p^β.
> If m is the number of distinct subgroups of G of order p^β, then
>
> $$m \equiv 1 \ (\mathrm{mod}\ p).$$

Our strategy for proving Theorem 5.1 is to examine the steps of the proofs of the first two Sylow theorems and to generalize them.

The core of the original proofs was a group action on the set of p^α-element subsets of the group. For the generalization we define the set X on which G acts by left multiplication to be the set of all p^β-element subsets of G.

As in the first two Sylow theorems, our proof will start by considering the size of X. We have

$$|X| = {}^nC_{p^\beta}.$$

Unlike in the proofs of the first two Sylow theorems, this binomial coefficient may well be divisible by p, as the next exercise illustrates.

Exercise 5.2

Let G be a group of order 24 and let X be the set of 4-element subsets of G.

(a) Calculate $|X|$.

(b) What is the highest power of 2 which divides $|X|$?

Let Y be the set of 2-element subsets of G.

(c) Calculate $|Y|$.

(d) What is the highest power of 2 which divides $|Y|$?

Exercise 5.2 gives a hint as to the general result. For the 2^2-element subsets, 2^1 was the highest power dividing $|X|$. For the 2^1-element subsets, the highest power dividing $|Y|$ was 2^2. In each case, the combination of the size of the subsets and the size of the collection of subsets gave us 2^3 altogether, accounting for all the factors of 2 in the order of G.

Thus, if we are considering the set X of p^β-element subsets, we might expect the highest power of p which divides

$$|X| = {}^nC_{p^\beta}.$$

to be $p^{\alpha-\beta}$. Once we establish this fact, the rest of the proof of Theorem 5.1 will follow those of the first two Sylow theorems exactly.

Proof of Theorem 5.1

Suppose that $|G| = p^\alpha z$ where p does not divide z and let $0 \leq \beta \leq \alpha$.

Let X be the set of p^β-element subsets of G and let G act on X by left multiplication.

> This is a group action; the previous proof carries over without change.

Now

$$|X| = {}^{p^\alpha z}C_{p^\beta} = \frac{(p^\alpha z - 0) \times (p^\alpha z - 1) \times \cdots \times (p^\alpha z - (p^\beta - 1))}{(p^\beta - 0) \times (p^\beta - 1) \times \cdots \times 1}$$

$$= \frac{(p^\alpha z - 0) \times (p^\alpha z - 1) \times \cdots \times (p^\alpha z - (p^\beta - 1))}{(p^\beta - 0) \times (p^\beta - 1) \times \cdots \times (p^\beta - (p^\beta - 1))}.$$

Each term in the numerator is of the form
$$p^\alpha z - k,$$
where k runs from 0 to $p^\beta - 1$.

Equally, each corresponding term in the denominator is of the form
$$p^\beta - k,$$
where k runs from 0 to $p^\beta - 1$.

For all terms except the first, in either numerator or denominator, a term will be divisible by a given power of p if and only if k is divisible by that power of p. In each such case, the highest power of p dividing corresponding terms in the numerator and denominator will be precisely the same, namely the highest power which divides k.

You proved this in Exercise 2.1.

Once we have cancelled all these powers of p, the only remaining power of p in $^{p^\alpha z}C_{p^\beta}$ arises from the first terms, p^α in the numerator and p^β in the denominator. Hence the power of p dividing $|X|$ is
$$\frac{p^\alpha}{p^\beta} = p^{\alpha-\beta}$$
as claimed. Thus, we may write
$$|X| = p^{\alpha-\beta} y,$$
where y is not divisible by p.

Now, as with the first two Sylow theorems, we consider the partition equation for the action of G on X defined by left multiplication.

Suppose that the distinct orbits are
$$\mathrm{Orb}(A_1), \ldots, \mathrm{Orb}(A_s).$$
Then the partition equation gives
$$|X| = |\mathrm{Orb}(A_1)| + \cdots + |\mathrm{Orb}(A_s)|.$$
There must be a term on the right-hand side divisible by a power of p which is at most $p^{\alpha-\beta}$, because, if not, all would be divisible by a higher power, contradicting what we have just proved about $|X|$.

Let A be an orbit whose size is divisible by at most $p^{\alpha-\beta}$.
By the Orbit–stabilizer Theorem, we have
$$|G| = |\mathrm{Orb}(A)| \times |\mathrm{Stab}(A)|.$$
Consider the powers of p on both sides. On the left we have p^α. The first term on the right contributes *at most* $p^{\alpha-\beta}$. It follows that $|\mathrm{Stab}(A)|$ must contribute *at least* p^β. Hence
$$|\mathrm{Stab}(A)| \geq p^\beta.$$
As before, we now look at the connection between $\mathrm{Stab}(A)$ and A.
By Lemma 1.1,
$$|\mathrm{Stab}(A)| \text{ divides } |A|.$$
Hence
$$|\mathrm{Stab}(A)| \leq |A| = p^\beta.$$
This second inequality completes the proof that
$$|\mathrm{Stab}(A)| = p^\beta$$
and we have produced the subgroup $\mathrm{Stab}(A)$ of order p^β whose existence was claimed.

We have proved a little more on the way. We started by assuming that $|\mathrm{Orb}(A)|$ was divisible by *at most* $p^{\alpha-\beta}$. We then showed that $|\mathrm{Stab}(A)|$ was precisely p^β. The relation
$$|G| = |\mathrm{Orb}(A)| \times |\mathrm{Stab}(A)|.$$
now shows that $|\mathrm{Orb}(A)|$ is divisible by *precisely* $p^{\alpha-\beta}$. In fact, we have
$$|\mathrm{Orb}(A)| = p^{\alpha-\beta} z.$$

Also by Lemma 1.1, for any $A_i \in X$, $\text{Stab}(A_i)$ is a p-group, of order p^γ say ($0 \leq \gamma \leq \alpha$). Therefore, by the Orbit–stabilizer Theorem,

$$|\text{Orb}(A_i)| = p^{\alpha-\gamma}z,$$

and so the sizes of all orbits are divisible by z. Furthermore,

$$|\text{Stab}(A_i)| \text{ divides } |A_i|.$$

So $\gamma \leq \beta$, and hence $p^{\alpha-\gamma} \geq p^{\alpha-\beta}$.

Now assume that there are m subgroups of order p^β. As in the proof of Sylow's Second Theorem, each orbit whose size is divisible by precisely $p^{\alpha-\beta}$ contains exactly one subgroup of order p^β. Let these orbits be $\text{Orb}(A_1)$ up to $\text{Orb}(A_m)$. The remaining orbits have sizes divisible by z and a power of p greater than $p^{\alpha-\beta}$ and hence divisible by

$$p^{\alpha-\beta+1}z.$$

As in the Sylow proof, the orbits consist of cosets, only one of which can be a subgroup.

Gathering terms in the partition equation, we obtain

$$|X| = p^{\alpha-\beta}y = mp^{\alpha-\beta}z + (\text{terms divisible by } p^{\alpha-\beta+1}z).$$

Dividing by $p^{\alpha-\beta}$, we have

$$y = mz + kpz,$$

for some positive integer k which depends on the particular group of order $p^\alpha z$.

We now do the same as in the proof of Sylow's Second Theorem and apply this result to the cyclic group $\mathbb{Z}_{p^\alpha z}$, of order $p^\alpha z$, for which we know that the corresponding value of m is 1. That is,

$$y = 1 \times z + k'pz,$$

A cyclic group has a unique subgroup corresponding to each divisor of its order.

for some (other) positive integer k'.

Equating the two expressions for y gives

$$mz + kpz = z + k'pz.$$

Dividing by z gives

$$m + kp = 1 + k'p.$$

Lastly, reducing modulo p gives

$$m \equiv 1 \pmod{p}.$$

This completes the proof. ∎

Theorem 5.1 provides additional restrictions on the number of subgroups of a particular order that a group may possess. These restrictions may be useful in helping to classify groups of a particular order.

Although we cannot say that two subgroups of a particular prime power order must be conjugate, we can rescue something from Sylow's Third Theorem. Any conjugate of a subgroup of order p^β is also a subgroup of order p^β. If we can show that such a subgroup is unique ($m = 1$), then we can deduce that it is equal to all its conjugates, that is it is normal. However, in the absence of coprimeness, uniqueness is usually hard to prove.

Exercise 5.3

Show that any group of order 196 has a normal subgroup of order 49. Give an example of a group of order 196 which has more than one subgroup of order 7.

This concludes our investigation of subgroups of prime power orders. We shall, however, make some use of the various theorems in *Unit GR6*.

SOLUTIONS TO THE EXERCISES

Solution 1.1

Following the scheme in Example 1.1, let G be any group of order 20. Let X be the set of all 5-element subsets of G and define an action of G on X by

$$g \wedge A = gA,$$

for every element A of X.

The number of 5-element subsets of G is

$$\begin{aligned} {}^{20}C_5 &= \frac{20 \times 19 \times 18 \times 17 \times 16}{5 \times 4 \times 3 \times 2 \times 1} \\ &= \frac{4 \times 19 \times 6 \times 17 \times 16}{1 \times 4 \times 1 \times 2 \times 1} \\ &= 15504. \end{aligned}$$

Thus $|X| = 15504$.

In Example 1.1, the significant fact about $|X|$ was that it was *not* divisible by the prime being considered (2 in the example, 5 here). We note that, here, $|X| = 15504$ is not divisible by 5.

Suppose that

$$\mathrm{Orb}(A_1), \mathrm{Orb}(A_2), \ldots, \mathrm{Orb}(A_s)$$

are the distinct orbits for this group action. Then

$$15504 = |\mathrm{Orb}(A_1)| + |\mathrm{Orb}(A_2)| + \cdots + |\mathrm{Orb}(A_s)|.$$

If all terms on the right-hand side were divisible by 5, then the left-hand side would be divisible by 5. It follows that at least one term on the right-hand side is not divisible by 5.

Suppose that A is an element of X such that $|\mathrm{Orb}(A)|$ is not divisible by 5. The Orbit–stabilizer Theorem gives

$$20 = |G| = |\mathrm{Orb}(A)| \times |\mathrm{Stab}(A)|.$$

Since $|\mathrm{Orb}(A)|$ is not divisible by 5, $|\mathrm{Stab}(A)|$ must be. Hence

$$|\mathrm{Stab}(A)| \geq 5.$$

As in the example, we now derive the opposite inequality.
If g is any element of $\mathrm{Stab}(A)$ then

$$g \wedge A = gA = A.$$

Thus, for a fixed element a of A, we have $ga \in A$, for all elements g of $\mathrm{Stab}(A)$. Hence the whole of the right coset

$$\mathrm{Stab}(A)a \subseteq A.$$

Thus, because a right coset has the same number of elements as the subgroup from which it is formed,

$$|\mathrm{Stab}(A)| = |\mathrm{Stab}(A)a| \leq |A| = 5.$$

Hence $|\mathrm{Stab}(A)| = 5$ and $\mathrm{Stab}(A)$ is a subgroup of G of order 5.

Solution 1.2

(a) We have to show that, if A is a subset of G with k elements, then so is gA.

Since gA consists of products of elements of G, by closure it is a *subset* of G.

We show that gA has k elements by showing that the function from A to gA defined by
$$a \mapsto ga$$
is one–one and onto.

One–one Suppose that $ga_1 = ga_2$ then, by left cancellation, $a_1 = a_2$. Hence the function is one–one.

Onto This follows from the definition of gA, since every element is of the form ga, for some element a of A.

Thus A and gA have the same number of elements, k, and so $g \wedge A = gA \in X$.

(b) We have
$$e \wedge A = eA = \{ea : a \in A\} = \{a : a \in A\} = A.$$

(c) By the definition of $\wedge$ and associativity in G, if g and h are elements of G, we have
$$\begin{aligned}(gh) \wedge A &= (gh)A \\ &= \{(gh)a : a \in A\} \\ &= \{g(ha) : a \in A\} \\ &= g \wedge \{ha : a \in A\} \\ &= g \wedge (hA) \\ &= g \wedge (h \wedge A).\end{aligned}$$

Solution 1.3

(a) The highest power of 3 which divides 24 is $3^1 = 3$. Also
$$\begin{aligned}{}^{24}C_3 &= \frac{24 \times 23 \times 22}{3 \times 2 \times 1} \\ &= \frac{8 \times 23 \times 22}{1 \times 2 \times 1} \\ &= 8 \times 23 \times 11 \; (= 2024)\end{aligned}$$
which is not divisible by 3. This fills in the gap in the generalization of Example 1.1, and we may deduce that any group of order 24 has a subgroup of order 3.

(b) The highest power of 2 dividing 24 is $2^3 = 8$. We have
$$\begin{aligned}{}^{24}C_8 &= \frac{24 \times 23 \times 22 \times 21 \times 20 \times 19 \times 18 \times 17}{8 \times 7 \times 6 \times 5 \times 4 \times 3 \times 2 \times 1} \\ &= \frac{3 \times 23 \times 11 \times 21 \times 5 \times 19 \times 9 \times 17}{1 \times 7 \times 3 \times 5 \times 1 \times 3 \times 1 \times 1} \\ &= 23 \times 11 \times 3 \times 19 \times 3 \times 17 \; (= 735471)\end{aligned}$$
which is not divisible by 2. As above, we may deduce that every group of order 24 has a subgroup of order 8.

The details of the calculations set out here give some indication of *why* the binomial coefficient ${}^{24}C_3$ is not divisible by 3 and why the binomial coefficient ${}^{24}C_8$ is not divisible by 2. In the first case the only term in the numerator divisible by 3 is 24, and its factor of 3 cancels with the corresponding term 3 in the denominator. In the second case every term in the original numerator divisible by a power of 2 has a corresponding term in the denominator divisible by the *same* power of 2.

Solution 1.4

The calculations are slightly shorter here.

$$^{15}C_3 = \frac{15 \times 14 \times 13}{3 \times 2 \times 1}$$
$$= \frac{5 \times 7 \times 13}{1 \times 1 \times 1}$$
$$= 5 \times 7 \times 13 \ (= 455)$$

Notice that the cancellations occur in corresponding terms as in the previous solution.

which is not divisible by 3. Hence G has a subgroup H of order 3.

$$^{15}C_5 = \frac{15 \times 14 \times 13 \times 12 \times 11}{5 \times 4 \times 3 \times 2 \times 1}$$
$$= \frac{3 \times 7 \times 13 \times 6 \times 11}{1 \times 2 \times 3 \times 1 \times 1}$$
$$= 3 \times 7 \times 13 \times 11 \ (= 3003)$$

which is not divisible by 5. Hence G has a subgroup K of order 5.

Solution 1.5

(a) The subgroup which is the intersection of two Sylow 5-subgroups must, by Lagrange's Theorem, have order 1 or 5. Since the subgroups are *distinct*, the intersection must have order 1 and hence be trivial.

(b) There are m subgroups of order 5, where m is congruent to 1 modulo 5. The possibilities for m are $1, 6, \ldots$. However, by the result just proved, every new 5-element subgroup contributes 4 more non-identity elements of order 5. Thus, if there were 6 or more such subgroups, we would have the identity plus at least 24 other elements. Since this is impossible, we conclude that $m = 1$.

This unique Sylow 5-subgroup contains the identity and 4 elements of order 5. Since any element of order 5 in G would generate a Sylow 5-subgroup, G can have only these 4 elements of order 5.

(c) By a similar argument to that for order 5, any two distinct subgroups of order 3 have trivial intersection. Each such subgroup has two non-identity elements of order 3. Hence there are $2l$ elements of order 3 in G.

(d) Elements of G can only have orders 1, 3, 5 and 15.

The identity is the only element of order 1.

There are 4 elements of order 5.

There are $2l$ elements of order 3.

If there are no elements of order 15 then we must have

$$15 = 1 + 4 + 2l$$

which gives $l = 5$. However 5 is not congruent to 1 modulo 3 as required by Sylow's Second Theorem. This contradiction shows that G must have an element of order 15 and is, therefore, cyclic.

Solution 2.1

(a) (i) Suppose that p^β divides $p^\alpha z - k$. Then
$$p^\alpha z - k = p^\beta l$$
for some integer l. Thus
$$k = p^\alpha z - p^\beta l.$$
Suppose now that $\beta \geq \alpha$, so that $p^\beta \geq p^\alpha$. Then
$$k = p^\alpha z - p^\beta l$$
$$= p^\alpha (z - p^{\beta - \alpha} l).$$
Thus p^α divides k, contradicting the fact that $0 < k < p^\alpha$. Hence $\beta < \alpha$ as required.

(ii) Again suppose that p^β divides $p^\alpha z - k$. Then, as above,
$$k = p^\alpha z - p^\beta l$$
for some integer l. By part (i), $\beta < \alpha$, so that $p^\alpha > p^\beta$. Therefore
$$k = p^\alpha z - p^\beta l$$
$$= p^\beta (p^{\alpha - \beta} z - l).$$
Thus p^β divides k.

Conversely, suppose that p^β divides k. Then, as $0 < k < p^\alpha$, we must have $\beta < \alpha$ and so p^β divides p^α. Hence p^β divides both terms in
$$p^\alpha z - k,$$
and therefore divides $p^\alpha z - k$.

(b) Consider the terms in the binomial coefficient:
$$\frac{(p^\alpha z - 0) \times (p^\alpha z - 1) \times \cdots \times (p^\alpha z - (p^\alpha - 1))}{(p^\alpha - 0) \times (p^\alpha - 1) \times \cdots \times (p^\alpha - (p^\alpha - 1))}.$$

The power of p dividing the first term in the numerator is clearly the same as the power of p dividing the first term in the denominator, namely p^α.

Each remaining term in the numerator is of the form $p^\alpha z - k$ with $0 < k < p^\alpha$. By part (a), the maximum power of p dividing $p^\alpha z - k$ is the maximum power of p dividing k. This is true for *all* integer values of z. In particular, it is true for the case $z = 1$. Hence, the maximum power of p dividing k is also the maximum power of p dividing $p^\alpha 1 - k = p^\alpha - k$, which is the term in the denominator corresponding to $p^\alpha z - k$ in the numerator.

Hence each factor of p in the numerator is cancelled out by a corresponding factor of p in the denominator.

Thus the binomial coefficient $^{p^\alpha z}C_{p^\alpha}$ is not divisible by p.

Solution 2.2

(a) Since
$$91 = 7 \times 13,$$
the Sylow 13-subgroups of G are of order 13 and hence cyclic. By an argument used in Solution 1.5, any two distinct subgroups of order 13 will have trivial intersection. Thus each new subgroup of order 13 contributes twelve additional elements of order 13.

If G has m Sylow 13-subgroups, then m is congruent to 1 modulo 13. Thus the possibilities for m are $m = 1, 14, \ldots$. But 14 such subgroups would give
$$1 + 14 \times 12 = 169$$
distinct elements, a contradiction since $|G| = 91$. Thus $m = 1$.

(b) So far, we know that G has one element of order 1 and twelve of order 13 (in the unique Sylow 13-subgroup). We now show that the remaining 78 elements cannot all be of order 7.

Since $91 = 7 \times 13$, the Sylow 7-subgroups of G must be cyclic of order 7. As before, any two distinct such subgroups have trivial intersection. Suppose that G has l Sylow 7-subgroups. Each such subgroup accounts for six elements of G of order 7.

Assume that there are no elements of order 91 (the only remaining possible order). Counting the elements of G, we have

$$91 = 1 + 12 + 6l,$$

which requires $l = 13$. But 13 is not congruent to 1 modulo 7, contradicting Sylow's Second Theorem.

Thus G has at least one element of order 91 and so is cyclic.

Solution 2.3

(a) Because $|G| = pq$, the Sylow q-subgroups of G are of prime order q and hence cyclic.

As in Solutions 1.5 and 2.2, each such subgroup has $q - 1$ elements of order q and any two distinct such subgroups have trivial intersection. So each new subgroup of order q contributes $q - 1$ additional elements of order q.

If G has m Sylow q-subgroups, then m is congruent to 1 modulo q and so the possibilities for m are $m = 1, q+1, \ldots$. However, $q+1$ such subgroups would contain

$$1 + (q+1)(q-1) = 1 + q^2 - 1 = q^2$$

elements. Since $p < q$, we have

$$|G| = pq < qq = q^2.$$

This contradiction shows that $m = 1$.

Sylow's Third Theorem, which we discuss in the next section, will provide a shorter proof that the subgroup is unique.

(b) The conjugate of a subgroup is a subgroup having the same order. Since the Sylow q-subgroup, H say, is unique, all conjugates of H must be H itself. That is,

$$gHg^{-1} = H, \quad \text{for all } g \in G,$$

and so H is normal in G.

Solution 3.3

(a) We know that l is congruent to 1 modulo 3 and that l divides 5. The first requirement gives

$$l = 1, 4, 7, \ldots$$

and the only one of these values satisfying the second restriction is $l = 1$.

(b) We know that m is congruent to 1 modulo 5 and that m divides 3. The first requirement gives

$$m = 1, 6, 11 \ldots$$

and the only one of these values satisfying the second restriction is $m = 1$.

(c) We now know that G contains the identity, just four elements of order 5 and just two elements of order 3. The remaining eight elements must, therefore, have order 15 (the only remaining possible order). Hence G is cyclic.

Solution 3.4

Suppose that G has l Sylow 3-subgroups. Then l is congruent to 1 modulo 3. So $l = 1, 4, 7, \ldots$, of which only 1 divides 2.
Hence G has a unique Sylow 3-subgroup.

Suppose that G has m Sylow 2-subgroups. Then m is congruent to 1 modulo 2. So $m = 1, 3, 5, \ldots$, of which only 1 and 3 divide 3.
Hence G could have one or three Sylow 3-subgroups.

There are only two different groups of order 6: $\mathbb{Z}_6$ and S_3.

Since $\mathbb{Z}_6$ is cyclic, it has a unique subgroup corresponding to each divisor of its order. Thus, $\mathbb{Z}_6$ has one Sylow 2-subgroup.

In S_3 each of the 2-cycles

$$(12), \quad (13) \quad \text{and} \quad (23)$$

generates a distinct subgroup of order 2. Hence S_3 has three Sylow 2-subgroups.

Solution 4.1

(a) Firstly, since $\mathbb{Z}_6$ is Abelian, there will be a normal (and hence unique) Sylow subgroup for each prime divisor.

Since $|\mathbb{Z}_6| = 2 \times 3$, the Sylow 2-subgroup will be of order 2, hence (isomorphic to) $\mathbb{Z}_2$. Similarly, the Sylow 3-subgroup will be $\mathbb{Z}_3$.

In terms of the elements of $\mathbb{Z}_6$, the subgroups are

$$\{0, 3\} \quad \text{and} \quad \{0, 2, 4\}.$$

Uniqueness is given by the normality together with the conjugacy part of Sylow's Third Theorem.

(b) Since the group is Abelian, the argument about uniqueness holds here too. The group has order

$$3 \times 15 = 3^2 \times 5.$$

We know that, since 3 and 5 are coprime, $\mathbb{Z}_{15} \cong \mathbb{Z}_3 \times \mathbb{Z}_5$, so

$$\mathbb{Z}_3 \times \mathbb{Z}_{15} \cong (\mathbb{Z}_3 \times \mathbb{Z}_3) \times \mathbb{Z}_5.$$

Thus the Sylow 3-subgroup (of order 3^2) is isomorphic to $\mathbb{Z}_3 \times \mathbb{Z}_3$ and the Sylow 5-subgroup (of order 5) is isomorphic to $\mathbb{Z}_5$.

As subgroups of the original group $\mathbb{Z}_3 \times \mathbb{Z}_{15}$, the Sylow 3-subgroup is

$$\mathbb{Z}_3 \times \{0, 5, 10\}$$

and the Sylow 5-subgroup is

$$\{0\} \times \{0, 3, 6, 9, 12\}.$$

This argument illustrates something more general, namely that the Sylow subgroups of a finite Abelian group are its primary components.

(c) The order is $6 = 2 \times 3$. Since S_3 is not Abelian, we cannot argue as above that the Sylow subgroups are unique.

However, the Sylow 3-subgroup is unique, for the following reasons. If there are m Sylow 3-subgroups, then we know that

$$m \equiv 1 \pmod{3} \quad \text{and} \quad m \mid 2.$$

The only solution is $m = 1$.

We can also derive this result from knowledge of the elements of S_3. Any Sylow 3-subgroup must be cyclic of order 3. There are only two elements of order 3, the 3-cycles

$$(123) \quad \text{and} \quad (132).$$

Thus the only possibility for a Sylow 3-subgroup is

$$\{e, (123), (132)\}.$$

Since this is a subgroup, it must be the unique Sylow 3-subgroup.

Any Sylow 2-subgroup must be cyclic of order 2. There are three elements of order 2, the 2-cycles

$$(12), \quad (13) \quad \text{and} \quad (23).$$

Each generates a cyclic subgroup of order 2, so the Sylow 2-subgroups are

$$\{e, (12)\}, \quad \{e, (13)\} \quad \text{and} \quad \{e, (23)\}.$$

The number of Sylow 2-subgroups is 3. As a check

$$3 \equiv 1 \,(\text{mod } 2) \quad \text{and} \quad 3 \mid 6.$$

As required by the theory, all the Sylow 2-subgroups are conjugate because their generators have the same cycle type and are, therefore, conjugate in S_3.

Solution 4.2

(a) There are several perfectly good proofs that D_7 has a unique Sylow 7-subgroup. All use the fact that, since the order of D_7 is $14 = 2 \times 7$, any Sylow 7-subgroup will have order 7.

First proof
Any Sylow 7-Subgroup must contain a generator of order 7. The only elements of order 7 are the non-trivial rotations. The rotation subgroup contains all such elements, has order 7 and so must be the unique Sylow 7-subgroup.

Second proof
Any Sylow 7-subgroup has index 2, is therefore normal and hence unique.

Third proof
The number of Sylow 7-subgroups is congruent to 1 modulo 7. Possible numbers are 1, 8, ... Each such subgroup would contribute 6 'new' elements of order 7, so 8 or more such subgroups are impossible. Thus the Sylow 7-subgroup is unique.

Fourth proof
The number of Sylow 7-subgroups is congruent to 1 modulo 7 and divides 2. The only possibility is 1.

(b) Any Sylow 2-subgroup of D_7 is of order 2 and hence cyclic. Again, there are several possible proofs that there are 7 such subgroups, we give two.

First proof
There are 7 reflections in D_7. Each generates a distinct subgroup of order 2, which is a Sylow 2-subgroup. There are no other elements of order 2.

Second proof
The number of Sylow 2-subgroups is congruent to 1 modulo 2, i.e is odd, and divides 7. The possibilities are 1 and 7.
Suppose there is a unique Sylow 2-subgroup. We count the elements of various orders.
There is only one element of order 1.
There are six elements of order 7 in the unique Sylow 7-subgroup.
By our assumption, there is only one element of order 2.
We have accounted for only eight elements of G. Hence the remaining elements must have order 14 (the only other possible order).
Hence

$$D_7 \cong \mathbb{Z}_{14},$$

a contradiction because $\mathbb{Z}_{14}$ is Abelian and D_7 is not.
The only remaining possibility is that there are 7 Sylow 2-subgroups.

(c) Each reflection is the non-identity element in a Sylow 2-subgroup. As all such subgroups are conjugate and the identity always conjugates to itself, the result follows.

(d) Using the notation of the presentation, the non-identity rotations are
$$a, a^2, a^3, a^4, a^5, a^6.$$

Conjugating a^i by another rotation, a^j say, gives
$$a^j a^i a^{-j} = a^{j+i-j} = a^i.$$

Conjugating by a reflection $a^j b$ gives
$$\begin{aligned}(a^j b)a^i(a^j b)^{-1} &= a^j(ba^i)b^{-1}a^{-j}\\ &= a^j a^{-i} bb^{-1} a^{-j} \quad \text{(since } ba^i = a^{-i}b\text{)}\\ &= a^{j-i-j}\\ &= a^{-i}.\end{aligned}$$

Thus every conjugate of a rotation is either itself or its inverse.

(e) The various calculations above show that the conjugacy classes in D_7 are
$$\{e\},$$
$$\{b, ab, a^2b, a^3b, a^4b, a^5b, a^6b\},$$
$$\{a, a^6\},$$
$$\{a^2, a^5\},$$
$$\{a^3, a^4\}.$$

The class equation is
$$14 = 1 + 7 + 2 + 2 + 2.$$

The centre consists of the single-element conjugacy classes. Hence the centre of D_7 is the trivial subgroup $\{e\}$.

Solution 4.3

We already know, from our work on Abelian groups, that there is only one Abelian group of order 14, namely $\mathbb{Z}_{14} \cong \mathbb{Z}_2 \times \mathbb{Z}_7$.

Let G be any group of order 14.

Using the Sylow theorems, there is a unique (normal) Sylow 7-subgroup, which must be cyclic of order 7. Let this subgroup be
$$\langle a : a^7 = e \rangle.$$

Again by the Sylow theorems, there is at least one Sylow 2-subgroup. Let
$$\langle b : b^2 = e \rangle$$
be one such subgroup.

In fact there are either 1 or 7 Sylow 2-subgroups.

Because b has order 2 it is not an element of $\langle a \rangle$. Therefore the cosets
$$\langle a \rangle = \{e, a, a^2, a^3, a^4, a^5, a^6\}$$
$$\langle a \rangle b = \{b, ab, a^2b, a^3b, a^4b, a^5b, a^6b\}$$
are distinct and give all 14 elements of G. In particular a and b generate G.

We now have most of the ingredients for a presentation of D_7. All that is left to do is to show that $ba = a^{-1}b$, or equivalently that $bab^{-1} = a^{-1}$.

The element bab^{-1} is a conjugate of a and so, as you can easily check, has order 7. It therefore belongs to the unique subgroup of order 7 generated by a. Hence
$$bab^{-1} = a^i \quad \text{for some integer } i,\ 0 < i < 7.$$

Conjugate elements always have the same order.

Note that, as bab^{-1} has order 7, it cannot be the identity, which excludes $i = 0$.

Conjugating again by b gives
$$\begin{aligned} b^2ab^{-2} &= b(bab^{-1})b^{-1} \\ &= ba^ib^{-1} \\ &= (bab^{-1})^i \\ &= (a^i)^i \\ &= a^{i^2}. \end{aligned}$$

Since $b^2 = e$, this gives
$$a = a^{i^2}.$$

So, as a has order 7, $i^2 \equiv 1 \pmod 7$ and, by exhaustion, $i = 1$ or 6.

The case $i = 1$ means that a and b commute. Since a and b generate G this implies that G is Abelian and hence
$$G \cong \mathbb{Z}_{14}.$$

We are left with the case $i = 6$, that is $bab^{-1} = a^6 = a^{-1}$. So in this case
$$G \cong \langle a, b : a^7 = e,\ b^2 = e,\ ba = a^6 b\, (= a^{-1}b) \rangle = D_7.$$

Solution 4.4

Since the order of D_{14} is $28 = 2^2 \times 7$, the Sylow subgroups will have orders 4 and 7. Any Sylow 7-subgroup must be cyclic. A Sylow 2-subgroup *could*, theoretically, be cyclic or the Klein group.

(a) The number of Sylow 7-subgroups must be 1, 8, ... and a divisor of 4. The only possibility is 1.

(b) The identity e has order 1.

The 14 reflections
$$b, ab, a^2 b, \ldots, a^{13} b$$
all have order 2.

The rotations form the subgroup
$$\{e, a, a^2, \ldots, a^{13}\},$$
of order 14. Thus, the only possible orders of the non-identity rotations are 2, 7 and 14. From our knowledge of cyclic groups we know that each of these does occur. For example, a has order 14, a^2 has order 7 and a^7 has order 2.

In fact, the orders of the rotations are as follows:

  1: e;
  2: a^7;
  7: $a^2, a^4, a^6, a^8, a^{10}, a^{12}$;
  14: $a, a^3, a^5, a^9, a^{11}, a^{13}$.

So, a^7 is the only rotation of order 2.

(c) The most straightforward argument is the 'parity' one. Since each reflection changes the colour of a 14-gon, their product preserves the colour and so is a rotation. Note that this includes the identity rotation, which is produced just in those cases where the two reflections are identical.

Any Sylow 2-subgroup is of order 4 and so can contain only elements of orders 1, 2 and 4. There are no elements of order 4, so each Sylow 2-subgroup contains the identity and three elements of order 2 (so is the Klein group).

There is only one rotation of order 2. Thus at least two elements of a Sylow 2-subgroup are reflections. However, the product of two such *distinct* reflections is a non-identity rotation. Thus each Sylow 2-subgroups contains two distinct reflections and the unique rotation of order 2.

(d) Conjugating a rotation must give a rotation (by the parity argument). Conjugation gives an element having the same order as the original. Hence, conjugates of a^7 must be rotations of order 2. There is only one such: a^7 itself. Hence, by Theorem 4.4 of *Unit GR4*,

$$\{e, a^7\}$$

is a normal subgroup.

We have already shown, in part (c), that a^7 (and hence $\{e, a^7\}$) lies in every Sylow 2-subgroup.

(e) The number of Sylow 2-subgroups is odd (congruent to 1 modulo 2) and divides 7. The possibilities are, therefore, only 1 and 7.

There are several ways to rule out 1, the easiest of which is as follows.

Each Sylow 2-subgroup contains e, a^7 and two 'new' reflections. There are, in all, 14 reflections (of order 2). To account for them requires 7 Sylow 2-subgroups.

Solution 4.5

(a) We have

$$|G| = 221 = 13 \times 17.$$

The number of Sylow 13-subgroups is congruent to 1 modulo 13 and divides 17. Thus there is a unique Sylow 13-subgroup.
The number of Sylow 17-subgroups is congruent to 1 modulo 17 and divides 13. Thus there is a unique Sylow 17-subgroup.

Thus G has unique, normal, cyclic Sylow subgroups of coprime orders 13 and 17. Hence, by Theorem 4.1 (and Theorem 5.1 of *Unit GR2*),

$$G \cong \mathbb{Z}_{13} \times \mathbb{Z}_{17} \cong \mathbb{Z}_{221},$$

and G is cyclic.

(b) We have $111 = 3 \times 37$. An argument like that above shows that a group of order 111 must have a unique Sylow 37-subgroup. However, the number of Sylow 3-subgroups can be either 1 or 37 since both are congruent to 1 modulo 3 and both divide 37. We therefore cannot be sure that there is a unique Sylow 3-subgroup.

Solution 4.6

The order of G has prime decomposition

$$110 = 2 \times 5 \times 11.$$

We check the number of Sylow subgroups corresponding to 2, 5 and 11 in turn.
For the 2-subgroups the number must be congruent to 1 modulo 2 and divide 55. There could thus be 1, 5, 11 or 55 such subgroups.
For the 5-subgroups the number must be congruent to 1 modulo 5 and divide 22. There could thus be 1 or 11 such subgroups.
For the 11-subgroups the number must be congruent to 1 modulo 11 and divide 10. There can only be one such subgroup.

Hence we can be sure that there is a unique Sylow 11-subgroup, which is normal because it is unique. Since it has order 11, this subgroup is non-trivial. Since its order is less than that of the group, it is proper.

Solution 4.7

The order factorizes into primes as

$$132 = 2^2 \times 3 \times 11.$$

Using the usual argument, the possibilities for the numbers of each type of Sylow subgroup are shown in the table.

Sylow 2-subgroups	1, 3, 11 or 33
Sylow 3-subgroups	1, 4 or 22
Sylow 11-subgroups	1 or 12

We want to rule out the possibility that there are more than one of every Sylow subgroup type. Suppose that we have 12 Sylow 11-subgroups, at least 4 Sylow 3-subgroups and at least 3 Sylow 2-subgroups.

Since the Sylow 11-subgroups are cyclic of prime order, the 12 subgroups account for the identity element plus 120 elements of order 11.
A similar argument shows that the 4 Sylow 3-subgroups, which are cyclic, account for 8 non-identity elements, each of order 3.

We have accounted for 129 elements of the group: the identity, 120 elements of order 11 and 8 elements of order 3. There are only 3 elements left, and these must be the 3 non-identity elements of a Sylow 2-subgroup (which has order 4). This contradicts the assumption that there are at least 3 Sylow 2-subgroups. Thus at least one Sylow subgroup must be unique and hence normal. Thus G cannot be simple.

Solution 5.1

(a) Since, from the presentation, a has order 4,

$$\langle a \rangle \cong \mathbb{Z}_4$$

is a cyclic subgroup of order 4.

(b) Closure is almost immediate. Probably the only combinations needing more than a mental check are the following:

$$\begin{aligned} ba^2 &= (a^3)^2 b \quad \text{(since } ba = a^3 b\text{)} \\ &= a^6 b \\ &= a^2 b; \end{aligned}$$

$$\begin{aligned} ba^2 b &= (a^2 b) b \quad \text{(by the above)} \\ &= a^2 b^2 \\ &= a^2; \end{aligned}$$

$$\begin{aligned} a^2 b a^2 &= a^2 (a^2 b) \quad \text{(by the above)} \\ &= a^4 b \\ &= b. \end{aligned}$$

The identity element is present and each element is self-inverse (see below). Thus we have a subgroup.

The element a^2 is a half-turn, the other two non-identity elements are reflections. Hence all non-identity elements are of order 2. The subgroup must be isomorphic to the Klein group.

Solution 5.2

(a) We have
$$|X| = \frac{24 \times 23 \times 22 \times 21}{4 \times 3 \times 2 \times 1}$$
$$= \frac{6 \times 23 \times 11 \times 21}{1 \times 3 \times 1 \times 1}$$
$$= 2 \times 23 \times 11 \times 21 \, (= 10626).$$

(b) From the factorized form above, it is clear that the highest power of 2 that divides $|X|$ is $2^1 = 2$.

(c) This time
$$|Y| = \frac{24 \times 23}{2 \times 1}$$
$$= 12 \times 23 \, (= 276).$$

(d) The highest power of 2 which divides $|Y|$ is 2^2.

Solution 5.3

We have $196 = 2^2 \times 7^2$. Thus the group has subgroups of order 7 and $7^2 = 49$.

Suppose that there are m subgroups of order 49. Then
$$m \equiv 1 \,(\mathrm{mod}\, 7) \quad \text{and} \quad m \mid 4.$$
Hence $m = 1$.
The Sylow 7-subgroup, of order 49, is unique and, hence, normal.

The Abelian group
$$\mathbb{Z}_4 \times \mathbb{Z}_7 \times \mathbb{Z}_7$$
has distinct subgroups
$$\{0\} \times \mathbb{Z}_7 \times \{0\} \quad \text{and} \quad \{0\} \times \{0\} \times \mathbb{Z}_7,$$
both of which have order 7.

OBJECTIVES

After you have studied this unit, you should be able to:

(a) apply the Sylow theorems to make deductions about the existence of subgroups;

(b) apply the Sylow theorems to the classification of groups of a particular order (in simple cases);

(c) apply counting arguments based on group actions to simple cases;

(d) follow the style of proof used in proving the Sylow theorems.

INDEX

binomial coefficient   7
internal direct product theorem   27
internal direct product theorem for finite groups   27
maximal prime power divisor   6
prime power subgroups theorem   30
simple group   25
Sylow p-subgroup   6, 13
Sylow's first theorem   6, 13, 14
Sylow's second theorem   12, 16
Sylow's third theorem   23
Sylow theorems   24